Messy Math

A Collection of Open-ended Investigations

How long would it take to write every number to a million?

Why is my birthday on a different day each year?

How many diapers will a baby use before he or she is toilet-trained?

Can a person really jump from the top of one building to the next like they do on TV?

Is it true that the length of your foot is the same as the distance between your elbow and your wrist?

How much water do I use while taking a shower?

Written by Paul Swan

Published by Didax Educational Resources®

www.didaxinc.com

Copyright © 2003 by Didax, Inc., Rowley, MA 01969. All rights reserved.

Limited reproduction permission: The publisher grants permission to individual teachers who have purchased this book to reproduce the blackline masters as needed for use with their own students. Reproduction for an entire school or school district or for commercial use is prohibited.

Printed in the United States of America.

This book is printed on recycled paper.

Order Number 2-161
ISBN 978-1-58324-159-2

E F G H I 15 14 13 12 11

395 Main Street
Rowley, MA 01969
www.didax.com

Foreword

So often the mathematics we present to students has been simplified to the point where all they need to do is follow a few memorized procedures in order to answer straightforward questions. Unfortunately, most questions we face in life are not straightforward and we need to apply our problem-solving skills in order to arrive at an answer—or one of many possible answers. Seldom do the problems we encounter contain all the information we require and so we need to seek out further information. At other times, we are faced with problems that contain too much information and we must determine which information is relevant.

This book is designed around a set of open-ended questions, many of which are drawn from magazines, newspapers and life experiences. As such, many of the questions involve sifting through information and collecting further information. To assist you, detailed notes providing extra hints and information have been provided. It is recommended, however, that you use discretion as to how and when this extra information is used. Let the students ask questions and seek out information before telling them too much. When they arrive at a solution or a possible set of solutions, encourage them to discuss the range of answers or the reasons for settling on a single answer. As students become experienced working with open-ended questions, you will find that they discover more questions that require answers and so on.

Contents

Teacher's Notes	4	Typing a Million	42–43
8,000 Diapers	6–7	A Feel for Big Numbers	44–45
The Pencil Case Problem—1	8–9	A Million Dots	46–47
The Pencil Case Problem—2	10–11	Large Number Lists	48–49
Tape Measure	12–13	Metric Paper—1	50–51
Take a Flying Leap	14–15	Metric Paper—2	52–53
Big Foot	16–17	Paper, Paper Everywhere	54–55
Down the Drain	18–19	My Legs Hurt	56–57
Pieces of Eight	20–21	Popcorn Packets	58–59
Competition Math	22–23	Measure Matters—1	60–61
Crossing the Road	24–25	Measure Matters—2	62–63
Tinkering with Time	26–27	Measure Matters—3	64–65
Metric Time	28–29	Measure Matters—4	66–67
Calendar Corrections	30–31	The Largest Organ in Your Body	68–69
The World Calendar	32–33	The Language of Math	70–71
The Calendar	34–35	Projections, Enlargements and Distortions	72–77
Four Calendars	36–37	Who Erased the Blackboard?	78–79
Calendar Conundrums	38–39	The Broken Calculator	80–81
Writing a Million	40–41	Where to Go from Here	82

Teacher's Notes

This book is designed around a set of explorative activities to help children:

- think
- talk
- research

in order to solve a problem.

Leading mathematics educator Paul Trafton believes that teachers should "make math messy" in order to encourage children to think in mathematics. When you consider it, most mathematics in real life is complicated. Either you don't possess enough information to solve a problem and you have to search for more, or you are given too much information and you need to discard some. Often in real life, there is not one right answer, but a variety of answers equally valid depending upon the circumstances. To illustrate, consider all the different cell phone plans available—comparing and choosing the best one is rather complicated.

The availability of technology such as calculators, computers and the Internet have now made it possible to ask open-ended questions within a classroom setting. Students working on open-ended investigations will often need to use calculators because the numbers will come from real data. They will also need to ask clarifying questions, make assumptions, estimate, survey, locate information and generally follow the pattern that a real-life person would when solving open-ended questions.

As a teacher, I would encourage you to have fun exploring the questions with the students. This will require a level of risk-taking, however a wealth of support material has been provided to complement each activity. The support material often provides the data required to solve the problem, ideas for extending the problem, or simply background information and stories to help the problem come alive for the students. Where possible, historical and cultural links to the mathematics contained on the activity page have been provided so that a context for the problem may be given.

A word of caution—to gain the most from these open-ended investigations, each student will need to be given the time to complete this problem–sorting method:

- discuss the problem;
- ask questions;
- explore the problem;
- ask questions; and
- explain the problem (present his or her findings).

Expect and encourage students, to explore various aspects of the problem; some will even go off on a tangent to the problem (that is how you will find new mathematics to explore). Fire their imaginations, learn along with them and boldly go where few have gone before.

TEACHER BACKGROUND

8,000 Diapers

As a father of twins in diapers, this article in *Choice* magazine—(August, 1999) the official journal of the Australian Consumers' Association—caught my eye.

This question offers many opportunities for investigation beyond the collection of data. For example, consider the environmental impact of 8,000 disposable diapers multiplied by the number of children wearing diapers in a particular state or country.

This question allows students the opportunity to collect primary data. There are always mothers with young children in and around elementary schools. Students could design and administer a survey in order to collect data on which to make an estimate. Clearly, the figure of 8,000 diapers is an estimate. After sorting the data it should become clear that it is impossible to arrive at an exact number—there are too many. Some children are toilet-trained earlier than others, some use more diapers, some parents use cloth diapers and so on.

Further Investigations

This investigation should prompt the students to think of some other data collection exercises such as:
- How much paper is consumed in a school each year?
- How many photocopies are made in a year?

8,000 Diapers

Plan

? Questions

A *Choice* magazine article suggested that by the time a toddler is toilet-trained he or she will have used 8,000 diapers. Another estimate suggested a child would have used 7,000 diapers by the time he or she was toilet-trained.

Which estimate do you think is closer?

_____ diapers

Collect some data to answer the question.

Where will you find the information?

Hints and Ideas

By what age is the typical toddler toilet-trained?
I could ask some moms around the school.

How many diaper changes does a baby have each day?

What did you find out?

Notes and Calculations

Findings

How will you present your findings?

Teacher Background

The Pencil Case Problem—1

This problem arose because my wife and I chose to call our second son Leighland. Like all teachers, I had trouble coming up with a name that didn't send shivers down my spine every time it was spoken. When Leighland started first grade, we bought him a pencil case with letter slots for his name. Unfortunately, his name was made up of nine letters and there were only eight slots on the pencil case. This was never a problem for our eldest son, Jeremy. It is a fairly simple exercise to collect length of name data in order to check whether eight slots are enough.

Further Investigations

Use the Internet to contact a class from another country to compare the length of names.

The Pencil Case Problem—1

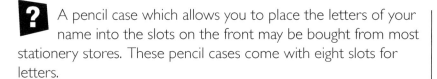

Questions

? A pencil case which allows you to place the letters of your name into the slots on the front may be bought from most stationery stores. These pencil cases come with eight slots for letters.

Collect some data to determine whether eight slots are enough for a student to fill in the:

- letters of his or her first name;
- letters of his or her last name and first initial;
- letters of his or her first name and the initial;
- letters of his or her nickname.

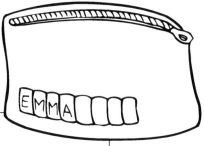

Where will you find the information?

Hints and Ideas

I could collect data from another class to check my findings.

My name is Alexander. It won't fit. I guess I could use Alex.

I'm going to list all the names of nine letters or more that I can think of.

What did you find out?

Notes and Calculations

Findings

How will you present your findings?

 You may wish to collect some data via the Internet to determine how the names of students from other countries would work.

Teacher Background

The Pencil Case Problem—2

This problem was tried with a class which contained quite an ethnic mix; hence, the second pencil case problem arose. The pencil cases come with a set of letters but when the letters are examined more closely, the unusual mix of letters becomes more apparent. For example there are only two A's, I's and O's and only one U. Vowels are used in most English words but perhaps the frequency differs in names. A class list provides an excellent opportunity for collecting data, as does a baby name book.

Further Investigations

The distribution of letters in English may be compared with the distribution in popular names. A look at the game of Scrabble™ also stimulates a great deal of discussion.

The Pencil Case Problem—2

 The same pencil case comes with a set of letters, shown below, which may be cut apart to fill the slots.

A	A	B	B	B	C	C	D	D	E	E	E	F	F
G	G	H	H	I	I	J	K	K	L	L	L	M	M
N	N	N	O	O	P	P	Q	R	R	S	S	S	
T	T	U	V	V	W	X	Y	Z	"	"	!	*	?

Some letters such as E, L, N, R and S appear more often than others.

Collect some data to determine whether this assortment of letters is appropriate for most students.

Questions

Where will you find the information?

 My name is Alana and there aren't enough letter A's for me to finish my name.

I wonder why there are only two A's, I's, O's and only one U when there are three B's, L's, R's and S's.

The most commonly used letter of the alphabet is "e" in the English language, but is this true for names?

 Yes | No

What did you find out?

 How are the letters distributed in a game of Scrabble™?

_____

_____

Findings

Notes and Calculations

How will you present your findings?

Teacher Background

Tape Measure

A visit to http://www.pitape.co.uk provides some interesting background to this problem. Any search of "Pi Tape" on the internet will bring up similar sites. This problem really tests whether the students understand that the circumference equals the product of diameter and π (c = πd) or whether they have just memorized the formula. If they understand that π is about 3 and there is a 3:1 relationship between the diameter and the circumference—producing a tape becomes a fairly simple matter. Roughly every 3 cm or, to be more precise, 3.14 cm, the tape needs to be marked to show a 1-cm diameter division.

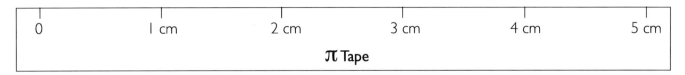

The finished tapes can then be tested on cylindrical objects where the diameter is already known. The need for accuracy may be emphasized by explaining that π Tapes are used in the engineering industry, particularly the aircraft and space industry, to accurately measure the diameters of cylindrical parts. The engineers, pilots and astronauts who fly the aircraft need to rely on the accuracy of these measuring devices. The topic of tolerances may also be discussed when producing the tapes.

Further Investigations

A related question involves the production of tennis ball cans. Some tennis ball manufacturers package three tennis balls per can. If you were to wrap a string around the can and another along the length of the can, which would be longer?

A similar investigation involves measuring the circumference and height of drinking glasses. In most cases, you will find the circumference exceeds the height measurement.

Messy Math

Tape Measure

Plan

? Design a tape measure that when wrapped around a cylinder could be used to directly read the diameter.

Questions

Hints and Ideas

The circumference of a circle is found by multiplying the diameter of the circle by π.

 The diameter runs through the middle of a circle.

π is a little more than three.

Where will you find the information?

 Where might a tape like this be used?

It could be used …
_____.

Who might use this type of tape?
_____

What did you find out?

Notes and Calculations

Findings

How will you present your findings?

© Didax Educational Resources® – www.didaxinc.com   13   Messy Math

Teacher Background

Take a Flying Leap

The record for men's long jump is around 8.95 m (approx. 29 ft) and for women, 7.5 m (approx. 24 ft). The students might like to measure their own attempts at the long jump, which would be significantly shorter than the world records. Buildings on the same street may be closer together than those across the street. Measuring the distance across a road near your school will provide a distance that may be used as an approximation in order to answer the question.

Take a Flying Leap

Plan

Questions

Last night on TV I saw a criminal running away from a police officer. They were running across the tops of buildings when they came to the end of the street. The criminal took a flying leap and just managed to make it to the building across the street. The offficer decided not to try. My dad said he was a wimp, but Mom said no one would be able to jump that far in real life.

Who do you think was right?

Hints and Ideas

How far can you jump? _____ m
or _____ ft

How far is it across an average street? _____ m
or _____ ft

I wonder what would happen if you jumped from a taller building onto a shorter building.

I wonder what the world record for long jump is.

Where will you find the information?

What did you find out?

Findings

Notes and Calculations

How will you present your findings?

Teacher Background

Big Foot

The footprint used on the page 17 belongs to one of my four-year-olds. Collecting foot sizes across several different age groups in the school should provide enough data to note several patterns. One clear pattern is that as a child ages and grows taller, his or her foot size increases. It is highly unlikely that a sixth grade student would have feet as small as the one shown on the page. The students should be able to suggest that the culprit was probably from kindergarten or first grade.

Further Investigations

Encourage each student to investigate other body measurements. He or she will need a piece of string as long as his or her arm span. Each student could be encouraged to explore the following relationships:

- Height and arm span – approximately the same
- Foot length and distance from elbow to wrist
- Twice around wrist and neck measurement
- Twice around neck and waist measurement
- Length of thumb and length of nose

Be sensitive about taking some body measurements—use discernment

The data from height and arm span can be used to draw a scattergraph and the relationship or correlation between the two variables may be examined. If correct, the points should be bunched and sloping around a line slanting up from the left to the right.

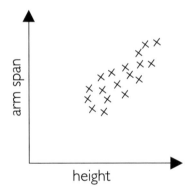

Big Foot

Plan

? Some sixth grade students were blamed for walking through wet cement. A copy of one of the footprints is shown below. The sixth graders are proclaiming their innocence! Put your detective caps on to find evidence to clear these students of any charges.

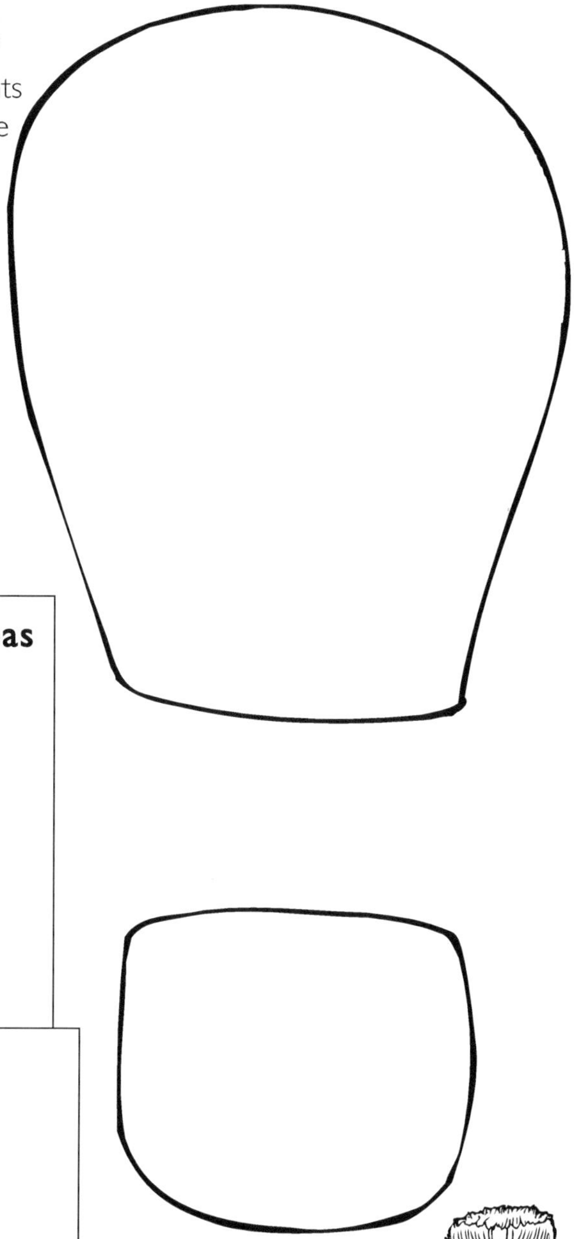

Questions

Where will you find the information?

Hints and Ideas

I heard that the length of your foot is the same as the distance between your elbow and wrist. I wonder if it is true.

Do taller people have larger feet?

I wonder what other body measurements are related.

What did you find out?

Findings

How will you present your findings?

Notes and Calculations

Messy Math

Teacher Background

Down the Drain

The following facts may add to your discussion:

- Showers use anywhere from 3 to 5 gallons of water every minute.
- Most toilets use 3.5 to 7 gallons of water per flush.
- A dual-flush toilet uses 60% less water.
- Leaving the water running while brushing your teeth wastes 2 gallons of water per minute.
- A dripping tap can waste 10 to 25 gallons of water per day.

Down the Drain

Plan

**? ** A great deal of water is wasted in the home. Did you know that every time you flush the toilet 3.5 to 7 gallons of water is used? A leaking tap may also waste thousands of gallons of water a year.

How much water do you think the average person uses while taking a shower?

_____ gal

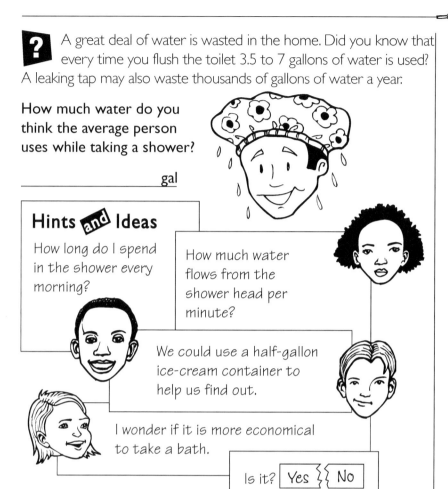

Hints and Ideas

How long do I spend in the shower every morning?

How much water flows from the shower head per minute?

We could use a half-gallon ice-cream container to help us find out.

I wonder if it is more economical to take a bath.

Is it? Yes / No

Notes and Calculations

Questions

Where will you find the information?

What did you find out?

Findings

How will you present your findings?

Messy Math

Teacher Background

Pieces of Eight

The most common form of currency used in the early American colonies was a "Spanish milled dollar" or "**piece of eight.**" England did not allow the American colonies to mint their own currency so they adopted the Spanish milled dollar.

The Spanish found precious metals such as gold and silver in Mexico and South America and began establishing mints for making coins. The first mint in America was established by the Spaniards in 1536 in Mexico City, followed soon after by one in South America. The most commonly minted coin was the Spanish milled dollar. The Spanish milled dollar was worth eight reales and was often cut up into eight pieces—hence the expression "pieces of eight." The number 8 was marked on the coin.

The Spanish milled dollar was made of one ounce of silver and had a patterned, or "milled," edge to prevent shaving the edges. Dividing the dollar into eight parts allowed various weights of silver to be produced; e.g. half an ounce, quarter of an ounce and eighth of an ounce. Americans to this day still use the terms "half dollar" and "quarter."

Pirates were always eager to plunder pieces of eight because they were popular in international exchange. The last Spanish milled dollars were minted in 1825 and they continued to be used as legal tender in the U.S. until 1857. There is evidence that they circulated into the early part of the twentieth century.

A **gold doubloon** was worth sixteen pieces of eight.

Pieces of Eight

? Last night I watched a pirate movie. The captain was scared that his crew might steal his treasure. While the crew was asleep, he carried his treasure chest full of pieces of eight (silver pieces) to the other side of the island he was on and buried it. My sister says there is no way that a person could carry that much on his own. I said pirates were strong.
Who was right?

Hints and Ideas

I wonder how large the treasure chest was.

I wonder how many pieces of eight would fit into the treasure chest.

Maybe we could try $1 coins to test our ideas.

Could the pirate carry the chest if it was filled with gold doubloons?

What were gold doubloons?

How much did a piece of eight weigh?

Notes and Calculations

Questions

Where will you find the information?

What did you find out?

Findings

How will you present your findings?

TEACHER BACKGROUND

Competition Math

The solution to the CD problem involves measuring the height of a CD in its cover and measuring the heights of various members of the class. Students can find an **average** or **mean** CD price to determine the value of the award. Many of the same techniques may be used to investigate the video/DVD option. Obviously a video option would be cheaper because, when packed in covers, a video is much thicker than a DVD. The worst case scenario would involve an extremely tall person (students could research this in the *Guinness Book of World Records*™) winning his or her height in DVDs.

The problem with running a similar competition using books is that the thickness of different books varies.

Further Investigations

Students can research the tallest man who ever lived by looking at the following websites:

http://www.altonweb.com/history/wadlow/

http://www.roadsideamerica.com/attract/ILALTwadlow.html

Note: Discussion about the CD competition might be stimulated by bringing in a stack of CDs and asking the students to estimate the worth of the collection.

Competition Math

? Radio stations often run competitions where the prize involves winning a stack of CDs the same height as the winner.

Hints and Ideas

How many CDs would you get if you won?
_____

How thick is a CD?
_____

How much do CDs cost?
_____

How much would the prize be worth?
_____

Notes and Calculations

The local video shop is thinking about running a similar competition using videos, DVDs, or perhaps a mix of each. Write a report outlining your suggestions for the competition. You might like to consider various options and the worst case scenario (for example, if a tall person won).

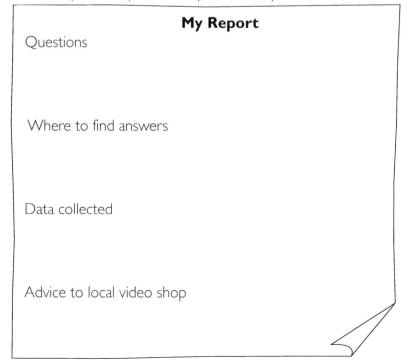

My Report

Questions

Where to find answers

Data collected

Advice to local video shop

The local bookstore is also considering a similar competition where the winner would receive a stack of books to match his or her height. What problems might occur if books were the prizes?
_____
_____

Teacher
Background
Crossing the Road

Most of us have been caught short crossing the road at traffic lights when the "don't walk" sign starts flashing and we are only halfway across the road.

Students will need to collect data such as:

- What is the width of a two lane and four lane road?
- How fast does the average person walk?
- How long does the signal allow for crossing the road?

A letter to the local city council will provide data on crossing sign times.

Crossing the Road

Plan

? Some traffic signals include walk signs which indicate when it is safe to cross the road. The walk signs are set on a timer which often makes a clicking noise prior to flashing a warning.

Collect data to determine how much time should be allowed for the average person to cross the road.

Hints and Ideas

How wide is a road with traffic lights?

What about older people or people with strollers and little children?

How fast does a person walk?

Notes and Calculations

Questions

Where will you find the information?

What did you find out?

Findings

How will you present your findings?

TEACHER BACKGROUND

Tinkering with Time

Much of our time work with students involves learning how to read the time from an analog clock. This activity focuses on the **passage of time**. Students' concept of time passing is affected by events, such as waiting for the day of a party to come (time seems to drag). Parents often use the expression "wait a sec" or "I'll only be 5 minutes"—and half an hour later a child is still waiting.

The study of **horology** is fascinating and there are many websites based on the development of accurate measuring devices for time. If children are given the opportunity to invent a device for measuring a set amount of time they will gain a better appreciation for the struggle to measure time. Nowadays we take clocks and watches for granted.

Hints and Ideas

There is a wealth of information to be found on the Internet under horology, clepsydra (water clock), sundial, hourglass and so on. The pendulum, still found in clocks today, was initially discovered by Galileo when he was bored in church one day. He noticed a lamp swinging in the cathedral and began to time the swings using his pulse as the measure. He discovered that the time it took to go back and forth was the same regardless of the size of the swing.

When experimenting with a pendulum, children should come to the realization that the period of the pendulum is independent of the mass of the weight on the end. However, changing the length of the string will affect the period of the pendulum.

Tinkering with Time

Plan

Many people have developed ways of estimating the passage of time. For example, to estimate how many seconds have passed, some people count 1,000 and 2,000 and 3,000 and 4,000 and so on, depending on the number of seconds. Throughout the ages, various devices have been used to measure time. For example, time has been measured using candle clocks, sand timers and water clocks. Use some of the materials listed below to make a device that measures a set number of seconds.

Materials List

plastic drink containers, plastic cup, rice, sand, marbles, plastic pipe, candle, string, washers and weights

Hints and Ideas

I could research the use of water, sand and candle clocks to measure time.

What is a pendulum?

You can make a pendulum using a string and a weight.

I count one elephant, two elephants, three elephants and so on.

A famous mathematician called Galileo investigated the use of pendulums to measure time.

Some clocks have a pendulum which swings below the clock.

What happens when you vary the length of the string? Does a heavier or lighter weight affect the pendulum?

My diagram

How it works

Does it work? Yes / No

Notes and Calculations

Teacher Background

Metric Time

With the advent of hundredths of a second, **milliseconds** and even **nanoseconds**, one could be forgiven for thinking that time was already based on powers of ten. Our system of measuring time appears to have begun thousands of years ago with the Sumerians who used a counting system based on 60, hence 60 seconds in a minute and 60 minutes in an hour. Several ancient societies used a **duodecimal system** (one based on 12–we still use a dozen when buying certain items like eggs) and hence 12 became significant in the keeping of time. At one time there were only 12 hours in a day (longer than the hour used today).

Further Investigations

For further background reading on the development of time see:

Waugh, A. *Time: From Micro-seconds to Millenia – A Search for the Right Time.* Headline Publishing, London, 1999.

Metric Time

Plan

? Most countries use the metric system for measurement. The units in this system are all related by multiples of ten. For example, there are 1,000 mm in a meter and 100 cm in a meter.

When it comes to measuring time, however, the system is a mess. Twenty-four hours make one day, there are 60 minutes in one hour and so on.

Design a metric system for measuring time.

Questions

Hints and Ideas

How did they come up with the metric system?

I heard it came from France.

One book I read said the standard for the meter was changed in the 20th century. I wonder if that is true.

Where will you find the information?

Notes and Calculations

What did you find out?

Findings

How will you present your findings?

Teacher Background

Calendar Corrections

The calendar has undergone many changes and corrections over the centuries. One of the most dramatic was the change from the **Julian Calendar** to the **Gregorian Calendar** in 1582. The Gregorian Calendar, which is still currently in use, includes the concept of a **leap year**, which the Julian Calendar did not. A leap year is required because a year is made up of 365.24 days, so every fourth year is given an extra day to compensate for the 0.24 extra part each year. Note: 4 x 0.24 does not equal a whole day, hence the rule that a leap year does not occur at the turn of the century unless the century is divisible by four without leaving a remainder; e.g. 1600, 2000. Note: 1800 and 1900 were not leap years.

The Julian Calendar was replaced by the Gregorian Calendar on October 4, 1582, in Catholic countries. People went to bed on October 4, and woke up on October 15. Many non-Catholic countries chose not to change their calendars. It wasn't until the 20th century that the final few countries changed over. Even today, some history books will list two dates for particular historical events—one date using the Julian Calendar and the other using the Gregorian Calendar.

For further information, see page 34.

Calendar Corrections

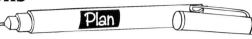

? Have you ever wondered why the tenth month of the year is called October when *Oct* means eight? Most people have to recite a little poem to remember the number of days in a particular month. What a mess!

Redesign the calendar to make it more simple to use.

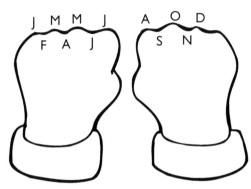

Hints and Ideas

30 days has September, April, June and November. All the rest have 31, except February. I wonder why some months have 31 days and poor old February only has 28.

If there are 52 weeks in a year and 7 days in a week, why are there 365 days in a year and not 364?

I use my knuckles to work out how many days in each month.

A leap year is used to keep the calendar in order. Why was the year 2000 a leap year but 1900 was not?

Notes and Calculations

Questions

Where will you find the information?

What did you find out?

Findings

How will you present your findings?

Teacher Background

The World Calendar

The upheaval caused by another change in the calendar would be so great that people the world over have resisted the temptation to change, even though several proposals have been suggested. Probably the most interesting proposal is:

The World Calendar (shown on page 33)

A search of the World Wide Web under "World Calendar" will bring up some interesting background information. You may like to show a copy to your students to see whether they can list any advantages and disadvantages to adopting this calendar.

The calendar affects our lives in many different ways and students need to gain an appreciation of the role that the calendar plays in society. Students should also realize that the Gregorian Calendar is not the only one in use. For example, consider the Jewish Calendar, the Chinese Calendar and the Buddhist Calendar. Apparently the Balinese operate two calendars besides the Gregorian Calendar, one calendar is used specifically for religious festivals.

Ask the students to explain what they notice about the World Calendar (i.e. any patterns).

The World Calendar

First Quarter

January
S	M	T	W	T	F	S
1	2	3	4	5	6	7
8	9	10	11	12	13	14
15	16	17	18	19	20	21
22	23	24	25	26	27	28
29	30	31				

February
S	M	T	W	T	F	S
				1	2	3	4
5	6	7	8	9	10	11
12	13	14	15	16	17	18
19	20	21	22	23	24	25
26	27	28	29	30		

March
S	M	T	W	T	F	S
					1	2
3	4	5	6	7	8	9
10	11	12	13	14	15	16
17	18	19	20	21	22	23
24	25	26	27	28	29	30

Second Quarter

April
S	M	T	W	T	F	S
1	2	3	4	5	6	7
8	9	10	11	12	13	14
15	16	17	18	19	20	21
22	23	24	25	26	27	28
29	30	31				

May
S	M	T	W	T	F	S
			1	2	3	4
5	6	7	8	9	10	11
12	13	14	15	16	17	18
19	20	21	22	23	24	25
26	27	28	29	30		

June
S	M	T	W	T	F	S
					1	2
3	4	5	6	7	8	9
10	11	12	13	14	15	16
17	18	19	20	21	22	23
24	25	26	27	28	29	30 W

Third Quarter

July
S	M	T	W	T	F	S
1	2	3	4	5	6	7
8	9	10	11	12	13	14
15	16	17	18	19	20	21
22	23	24	25	26	27	28
29	30	31				

August
S	M	T	W	T	F	S
			1	2	3	4
5	6	7	8	9	10	11
12	13	14	15	16	17	18
19	20	21	22	23	24	25
26	27	28	29	30		

September
S	M	T	W	T	F	S
					1	2
3	4	5	6	7	8	9
10	11	12	13	14	15	16
17	18	19	20	21	22	23
24	25	26	27	28	29	30

Fourth Quarter

October
S	M	T	W	T	F	S
1	2	3	4	5	6	7
8	9	10	11	12	13	14
15	16	17	18	19	20	21
22	23	24	25	26	27	28
29	30	31				

November
S	M	T	W	T	F	S
			1	2	3	4
5	6	7	8	9	10	11
12	13	14	15	16	17	18
19	20	21	22	23	24	25
26	27	28	29	30		

December
S	M	T	W	T	F	S
					1	2
3	4	5	6	7	8	9
10	11	12	13	14	15	16
17	18	19	20	21	22	23
24	25	26	27	28	29	30 W*

W = Leap Year Day, World Holiday (36th day), outside the week.

W* = Worldsday, World Holiday (365th day), outside the week.

Teacher Background

The Calendar

The calendar we currently use is called the **Gregorian Calendar**, dating back to Pope Gregory. It is an **annual calendar** because it changes each year. The reason it changes each year is because the 365 days that make up a year are not divisible by 7 without leaving a remainder. Note: When you divide 365 by 7 there is a remainder of one—hence each year starts and finishes on the same day and the next year will begin on the next day of the week. The overall consequence of this is that 7 different calendars are required to cover all possible starting days. A further 7 calendars are required to cope with leap years, which may begin on any day of the week. In total, 14 calendars are needed to cover all possibilities.

Four consecutive calendar years are shown on page 37.

The Calendar

 Plan

? My mother has a dish towel with last year's calendar printed on it. When I was washing up, I compared the calendar for last year with the new calendar for this year. I noticed that my birthday is on a different day this year. Last year my birthday was on a Wednesday; this year my birthday is on a Thursday. Looking at the calendar printed on the dish towel I began to wonder…

Our calendar is an annual calendar. It changes every year. When can we be able to use the same calendar again?

Questions

Where will you find the information?

Hints and Ideas

I will look at the calendars for four consecutive years.

On which day does this year begin and end?

I wonder what happens in a leap year.

What did you find out?

How many different calendars would you need to produce to cover all possibilities?

Hints and Ideas

What happens to your birthday in a leap year?

Does it matter whether your birthday falls before or after February 29?

I wonder what happens if you were born on February 29.

Findings

 **Notes and Calculations**

How will you present your findings?

 A character named Frederic in *The Pirates of Penzance* by Gilbert and Sullivan turned 21 after only 5 birthdays. Explain how this happened.

© Didax Educational Resources® – www.didaxinc.com   **Messy Math**

Teacher Background

Four Calendars

Used with the activity on page 37, **The Calendar** (see typed notes page 34).

Four Calendars

* 2004- is a leap year.

2002

January
M T W T F S S
　　1　2　3　4　5　6
7　8　9　10　11　12　13
14　15　16　17　18　19　20
21　22　23　24　25　26　27
28　29　30　31

February
M T W T F S S
　　　　　　1　2　3
4　5　6　7　8　9　10
11　12　13　14　15　16　17
18　19　20　21　22　23　24
25　26　27　28

March
M T W T F S S
　　　　　　1　2　3
4　5　6　7　8　9　10
11　12　13　14　15　16　17
18　19　20　21　22　23　24
25　26　27　28　29　30　31

April
M T W T F S S
1　2　3　4　5　6　7
8　9　10　11　12　13　14
15　16　17　18　19　20　21
22　23　24　25　26　27　28
29　30

May
M T W T F S S
　　　1　2　3　4　5
6　7　8　9　10　11　12
13　14　15　16　17　18　19
20　21　22　23　24　25　26
27　28　29　30　31

June
M T W T F S S
　　　　　　　　1　2
3　4　5　6　7　8　9
10　11　12　13　14　15　16
17　18　19　20　21　22　23
24　25　26　27　28　29　30

July
M T W T F S S
1　2　3　4　5　6　7
8　9　10　11　12　13　14
15　16　17　18　19　20　21
22　23　24　25　26　27　28
29　30　31

August
M T W T F S S
　　　　1　2　3　4
5　6　7　8　9　10　11
12　13　14　15　16　17　18
19　20　21　22　23　24　25
26　27　28　29　30　31

September
M T W T F S S
30　　　　　　　　1
2　3　4　5　6　7　8
9　10　11　12　13　14　15
16　17　18　19　20　21　22
23　24　25　26　27　28　29

October
M T W T F S S
　1　2　3　4　5　6
7　8　9　10　11　12　13
14　15　16　17　18　19　20
21　22　23　24　25　26　27
28　29　30　31

November
M T W T F S S
　　　　　1　2　3
4　5　6　7　8　9　10
11　12　13　14　15　16　17
18　19　20　21　22　23　24
25　26　27　28　29　30

December
M T W T F S S
30　31　　　　　　　1
2　3　4　5　6　7　8
9　10　11　12　13　14　15
16　17　18　19　20　21　22
23　24　25　26　27　28　29

2003

January
M T W T F S S
　　1　2　3　4　5
6　7　8　9　10　11　12
13　14　15　16　17　18　19
20　21　22　23　24　25　26
27　28　29　30　31

February
M T W T F S S
　　　　　　　1　2
3　4　5　6　7　8　9
10　11　12　13　14　15　16
17　18　19　20　21　22　23
24　25　26　27　28

March
M T W T F S S
31　　　　　　1　2
3　4　5　6　7　8　9
10　11　12　13　14　15　16
17　18　19　20　21　22　23
24　25　26　27　28　29　30

April
M T W T F S S
　1　2　3　4　5　6
7　8　9　10　11　12　13
14　15　16　17　18　19　20
21　22　23　24　25　26　27
28　29　30

May
M T W T F S S
　　　　1　2　3　4
5　6　7　8　9　10　11
12　13　14　15　16　17　18
19　20　21　22　23　24　25
26　27　28　29　30　31

June
M T W T F S S
30　　　　　　　　1
2　3　4　5　6　7　8
9　10　11　12　13　14　15
16　17　18　19　20　21　22
23　24　25　26　27　28

July
M T W T F S S
　1　2　3　4　5　6
7　8　9　10　11　12　13
14　15　16　17　18　19　20
21　22　23　24　25　26　27
28　29　30　31

August
M T W T F S S
　　　　　1　2　3
4　5　6　7　8　9　10
11　12　13　14　15　16　17
18　19　20　21　22　23　24
25　26　27　28　29　30　31

September
M T W T F S S
1　2　3　4　5　6　7
8　9　10　11　12　13　14
15　16　17　18　19　20　21
22　23　24　25　26　27　28
29　30

October
M T W T F S S
　　1　2　3　4　5
6　7　8　9　10　11　12
13　14　15　16　17　18　19
20　21　22　23　24　25　26
27　28　29　30　31

November
M T W T F S S
　　　　　　　1　2
3　4　5　6　7　8　9
10　11　12　13　14　15　16
17　18　19　20　21　22　23
24　25　26　27　28　29　30

December
M T W T F S S
1　2　3　4　5　6　7
8　9　10　11　12　13　14
15　16　17　18　19　20　21
22　23　24　25　26　27　28
29　30　31

2004*

January
M T W T F S S
　　　1　2　3　4
5　6　7　8　9　10　11
12　13　14　15　16　17　18
19　20　21　22　23　24　25
26　27　28　29　30　31

February
M T W T F S S
　　　　　　　　1
2　3　4　5　6　7　8
9　10　11　12　13　14　15
16　17　18　19　20　21　22
23　24　25　26　27　28　29

March
M T W T F S S
1　2　3　4　5　6　7
8　9　10　11　12　13　14
15　16　17　18　19　20　21
22　23　24　25　26　27　28
29　30　31

April
M T W T F S S
　　　1　2　3　4
5　6　7　8　9　10　11
12　13　14　15　16　17　18
19　20　21　22　23　24　25
26　27　28　29　30

May
M T W T F S S
31　　　　　1　2
3　4　5　6　7　8　9
10　11　12　13　14　15　16
17　18　19　20　21　22　23
24　25　26　27　28　29　30

June
M T W T F S S
1　2　3　4　5　6
7　8　9　10　11　12　13
14　15　16　17　18　19　20
21　22　23　24　25　26　27
28　29　30

July
M T W T F S S
　　　1　2　3　4
5　6　7　8　9　10　11
12　13　14　15　16　17　18
19　20　21　22　23　24　25
26　27　28　29　30　31

August
M T W T F S S
30　31　　　　　　1
2　3　4　5　6　7　8
9　10　11　12　13　14　15
16　17　18　19　20　21　22
23　24　25　26　27　28　29

September
M T W T F S S
　　1　2　3　4　5
6　7　8　9　10　11　12
13　14　15　16　17　18　19
20　21　22　23　24　25　26
27　28　29　30

October
M T W T F S S
　　　1　2　3
4　5　6　7　8　9　10
11　12　13　14　15　16　17
18　19　20　21　22　23　24
25　26　27　28　29　30　31

November
M T W T F S S
1　2　3　4　5　6　7
8　9　10　11　12　13　14
15　16　17　18　19　20　21
22　23　24　25　26　27　28
29　30

December
M T W T F S S
　　1　2　3　4　5
6　7　8　9　10　11　12
13　14　15　16　17　18　19
20　21　22　23　24　25　26
27　28　29　30　31

2005

January
M T W T F S S
31　　　　　　　1　2
3　4　5　6　7　8　9
10　11　12　13　14　15　16
17　18　19　20　21　22　23
24　25　26　27　28　29　30

February
M T W T F S S
　1　2　3　4　5　6
7　8　9　10　11　12　13
14　15　16　17　18　19　20
21　22　23　24　25　26　27
28

March
M T W T F S S
　1　2　3　4　5　6
7　8　9　10　11　12　13
14　15　16　17　18　19　20
21　22　23　24　25　26　27
28　29　30　31

April
M T W T F S S
　　　　1　2　3
4　5　6　7　8　9　10
11　12　13　14　15　16　17
18　19　20　21　22　23　24
25　26　27　28　29　30

May
M T W T F S S
30　31　　　　　　1
2　3　4　5　6　7　8
9　10　11　12　13　14　15
16　17　18　19　20　21　22
23　24　25　26　27　28　29

June
M T W T F S S
　　1　2　3　4　5
6　7　8　9　10　11　12
13　14　15　16　17　18　19
20　21　22　23　24　25　26
27　28　29　30

July
M T W T F S S
　　　1　2　3
4　5　6　7　8　9　10
11　12　13　14　15　16　17
18　19　20　21　22　23　24
25　26　27　28　29　30　31

August
M T W T F S S
1　2　3　4　5　6　7
8　9　10　11　12　13　14
15　16　17　18　19　20　21
22　23　24　25　26　27　28
29　30　31

September
M T W T F S S
　　　1　2　3　4
5　6　7　8　9　10　11
12　13　14　15　16　17　18
19　20　21　22　23　24　25
26　27　28　29　30

October
M T W T F S S
31　　　　　1　2
3　4　5　6　7　8　9
10　11　12　13　14　15　16
17　18　19　20　21　22　23
24　25　26　27　28　29　30

November
M T W T F S S
　1　2　3　4　5　6
7　8　9　10　11　12　13
14　15　16　17　18　19　20
21　22　23　24　25　26　27
28　29　30

December
M T W T F S S
　　　1　2　3　4
5　6　7　8　9　10　11
12　13　14　15　16　17　18
19　20　21　22　23　24　25
26　27　28　29　30　31

Messy Math

TEACHER
BACKGROUND
Calendar Conundrums

The library and the Internet contain a great deal of information about different cultures and the calendars they use. Many cultures use two calendars—the Gregorian, which most Western societies use, and their own. Most business is conducted according to the **Gregorian Calendar**, but religious festivals and celebrations are measured against traditional calendars. The Jewish and Buddhist calendars do not measure dates from the birth of Christ. The year number on the Jewish Calendar represents the number of years since the creation. This was determined by adding all the ages of the people in the Bible back to the beginning. The Jewish Calendar is a **lunar calendar**, each new month begins on the new moon. The problem with a lunar calendar is that the 12 months are too short (11 days are lost each year) and 13 months are too long (19 days are gained each year). Trying to match the lunar calendar and the Gregorian Calendar is the reason why the dates for certain religious ceremonies seem to change each year.

Most Buddhist calendars begin with Buddha's death, so the year 2002 is the year 2545 on the Buddhist Calendar.

Useful Facts

English Name	Roman Name	Meaning
January	Januarius	"of Janus," a two faced god looking back on and forward
February	Februarius	"of februa," a roman festival of purification held this month
March	Martius	"of mars" a roman god of war
April	Aprilis	to open- spring flowers bloom
May	Maius	"of Maia" meaning great one. Is the goddess of spring, growth
June	Junius	"of Juno" the goddess of marriage
July	Julius	named after Julius Caesar
August	Augustus	named after Augustus Caesar
September	September	roman for seven (remember, there was originally only ten months)
October	October	roman for eight
November	November	roman for nine
December	December	roman for ten

Note: Originally the year started in March and there were only ten months. January and February were added later. Each month consisted of 29 days and the year was only 290 days long!

The Jewish Calendar has the following months:

Name	Number	Length	Gregorian Equivalent
Nisan	1	30 days	March – April
Iyar	2	29 days	April – May
Sivan	3	30 days	May – June
Tammuz	4	29 days	June – July
Ab	5	30 days	July – August
Elul	6	29 days	August – September
Tishri	7	30 days	September – October
Heshvan	8	29 or 30 days	October – November
Kislev	9	30 or 29 days	November – December
Tebet	10	29 days	December – January
Shebat	11	30 days	January – February
Adar	12	29 or 30 days*	February – March

*In leap years, Adar has 30 days. In non-leap years, Adar has 29 days.

Calendar Conundrums

Plan

Questions

 Imagine life without a calendar. What would change in your life? What would stay the same?

Change

Stay the Same

Choose a research topic from the list below and present your findings to the class as a poster or report.

- What is the difference between a lunar and solar calendar?
- The Chinese Calendar
- The Jewish Calendar
- How the days were named
- How the months were named
- The Julian Calendar
- The Gregorian Calendar
- What do the abbreviations AM, PM, BC, AD, CE, BCE mean?
- Why does the date for Easter change each year?

Where will you find the information?

What did you find out?

Hints and Ideas

How would I know when the holidays start?

I could be in sixth grade forever without a calendar!

At least I wouldn't have to learn dates in history.

Did you know the year 46 BC was 455 days long?

The Egyptian year only had three seasons—the sowing, the growing and the flooding season!

There are many different calendars in use today besides the one we commonly use.

Findings

Did you know that in 1752, the start of the year was moved to January 1?

How will you present your findings?

Notes and Calculations

Teacher Background

Writing a Million

Mike Dolega finished writing the numbers from 1 to 1,000,000 on February 19, 1989. The Tasmanian man had spent 1,282 hours and used 97 ballpoint pens to fill the forty, 96-page note books. The process lasted over two years and has given Mike an appreciation of how many a million really is!

Further Investigations

The book, *Counting on Frank* (The Learning Company/Creative Wonders, 1994) by Rod Clement, includes plenty of opportunities for students to estimate and work with large numbers. One group of students, after reading the passage on the amount of ink in a ballpoint pen, tested the length of a line that could be drawn before the ink ran out. Consider how you would design such an experiment.

The Story of Gauss

Students could be told the story of Gauss—a famous mathematician who, at the age of 10, was punished for being naughty in class. The teacher told him to add all the numbers from 1 to 100. Most people would think that this punishment would take a long time to complete, so imagine the teacher's surprise when young Gauss finished the punishment in just a few minutes. Students should be given the opportunity to try the problem before explaining that he saw a pattern when the numbers were paired. The approach Gauss used may be illustrated using the numbers 1 to 10.

$$1 + 2 + 3 + 4 + 5 + 6 + 7 + 8 + 9 + 10$$

Each pair of numbers adds to 11 and there are five pairs, hence the total is 55.

Writing a Million

 nine hundred ninety-nine thousand, nine hundred ninety-seven; nine hundred ninety-nine thousand, nine hundred ninety-eight; nine hundred ninety-nine thousand, nine hundred ninety-nine; one million!

Imagine writing all the numbers from one to a million!

How many pages of paper do you think you would use?

Explain how you arrived at your estimate.

_____

_____

_____

_____

Hints and Ideas

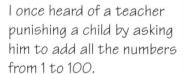
I think it would depend on whether the numbers were written as words or numerals.

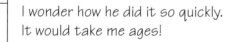
I once heard of a teacher punishing a child by asking him to add all the numbers from 1 to 100.

I heard he was able to add all the numbers from 1 to 100 in just a few seconds.

I wonder how he did it so quickly. It would take me ages!

Notes and Calculations

 You might be surprised to learn that a Tasmanian man, Mike Dolega, actually wrote all the number words from 1 to 1,000,000.

Plan

Questions

Where will you find the information?

What did you find out?

Findings

How will you present your findings?

Teacher Background

Typing a Million

Les Stewart from Mudjimba, Australia, is cited in the *Guinness Book of World Records* as holding the record for Typing Numbers in Word Form. He began typing in 1982 and finished sixteen years and seven months later. He managed to complete around three pages a day by typing 20 minutes on the hour, every hour.

Note: He typed using one finger and used seven typewriters and 1,000 ribbons. In total he used up 19,890 pages of paper. See **http://www.recordholders.org/en/records/typing.html**

To estimate how long this task would take, students need to find how long it takes to type numbers in word form. Obviously, it is fairly quick to type the word *"six"* but it would take a great deal more time to type *nine hundred ninety-nine thousand, nine hundred ninety-seven*. The students would need to type a variety of numbers to form an average typing time.

When calculating the number of pages required, students should use a plain font such as Courier, as this is closer to the original typewriter print. A picture of a typed page is shown on the website. This will give some indication of the spacing used.

Note: Having been brought up with computers and word processors, most students will not be familiar with mechanical or electrical typewriters, so they may not be aware of the reason behind the placement of keys on the standard "QWERTY" keyboard. Basically, the keys were placed this way in order to slow down typists who were jamming the mechanical keys on the typewriter because they were typing too quickly. In this day of word processors and computers, why is the same keyboard layout still used? Students might like to investigate alternative keyboard layouts by searching on the Internet.

Typing a Million

 nine hundred ninety-nine thousand, nine hundred ninety-seven; nine hundred ninety-nine thousand, nine hundred ninety-eight; nine hundred ninety-nine thousand, nine hundred ninety-nine; one million!

Imagine typing all the numbers from one to a million!

You might be surprised to learn that an Australian, Les Stewart from Mudjimba, holds the world record for typing numbers in word form. He used a manual typewriter and only one finger to press the keys.

Estimate how long you think this task would take. _____

Estimate how many sheets of paper would be used. _____

How did you arrive at your estimates? _____

Hints and Ideas

Sounds like a new type of punishment that a teacher thought of.

It would take years to type all those number words!

Remember you couldn't type all day every day.

How many lines can be typed onto a single page?

As the numbers become larger, there are more letters to type.

Notes and Calculations

Plan

Questions

Where will you find the information?

What did you find out?

Findings

How will you present your findings?

Teacher Background

A Feel for Big Numbers

Children are often fascinated with large numbers. This activity, along with **A Million Dots,** is designed to capitalize on this interest. News reports often refer to numbers in the billions and sometimes trillions. Yet, these numbers are meaningless unless we have some idea of their size.

The question, "Have you lived for a million seconds?" is designed to show how a million "small things" soon add up.

$1{,}000{,}000 \div 60 \div 60 \div 24$ shows that one million seconds is slightly over $11\frac{1}{2}$ days (11.57 days). Students may like to calculate the time in days, hours and minutes. Even 1,000,000,000 (one billion) seconds has passed by the time a person reaches 32 (31.7 yrs).

Counting from one to a million would take longer than $11\frac{1}{2}$ days because as the numbers become larger it would take longer, than one second to say each number. Students may also bring up fatigue, losing count and so on as possible difficulties.

See also page 49, **A Million Dots** and page 51, **Large Number Lists.**

A Feel for Big Numbers

Plan

Questions

? Did you know that it is easier to become a billionaire in the United States than it is in the United Kingdom?

In the United States a billion is 1,000,000,000 but in the United Kingdom and Australia 1,000,000,000 is only a thousand million. A billion in the United Kingdom is 1,000,000,000,000.

Hints and Ideas

I'm confused. How much money do you need to be called a billionaire in Australia?

Find out if you have lived for a million seconds.

- 60 minutes in an hour ...
- 24 hours in a day!
- 60 seconds in a minute ...

Where will you find the information?

Notes and Calculations

Yes / No

What did you find out?

How long would it take to count from one to a million?

Notes and Calculations

Findings

How will you present your findings?

Teacher Background

A Million Dots

This page is included to give students an idea of what a million looks like. The size of a billion starts to come into focus when you realize 100,000 of these pages would be required to print 1,000,000,000 dots.

A Million Dots

You will need 100 of these pages to make one million dots.

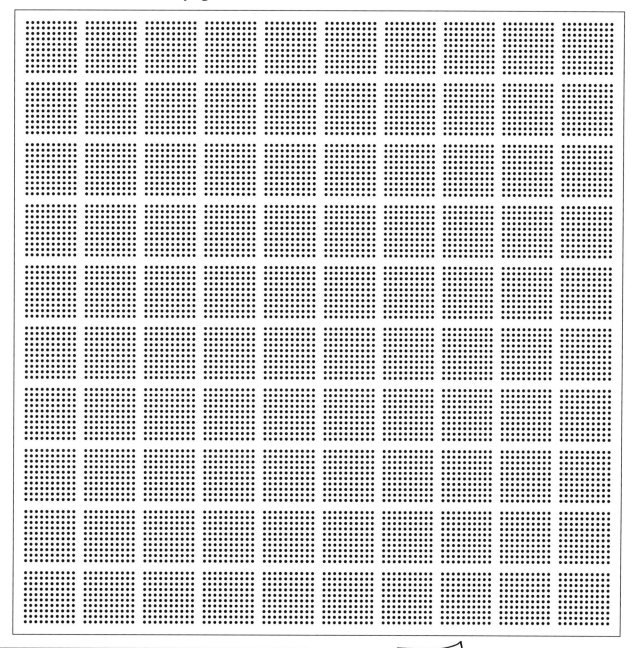

Hints and Ideas

Wow! Do you realize you would need 100,000 of these pages to make a US billion?

I wonder how long it would take if everyone in the school collected a million of the same thing.

How would you organize the counting?

How many pages would it take to make a googol?

How much space would it take up?

What's a googol?

I am going to collect one million of something—bread ties, bottle tops, ring pulls from cans…

© Didax Educational Resources® – www.didaxinc.com

Messy Math

Teacher Background

Large Number Lists

A page containing large numbers is included so students can see the pattern for naming numbers and so the size of numbers written as numerals can be observed. Students may also enjoy the story of how the **googol** was named. A quick look at the American large number chart shows that the names occur each time the numbers increase one thousand times or by three zeroes. Hence, a googol is an unusual number in terms of the naming convention.

1×10^{99}—one followed by 99 zeroes is called a **duotrigintillion** (US) or **sexdecilliard** (European).

1×10^{102}—one followed by 102 zeroes is called a **tretrigintillion** (US) or **septendecillion** (European). The googol fits between these two numbers and does not follow the standard naming convention.

A googol is 10,000,000,000,000,000,000,000,000,000,000,000,000,000,000,000,000,000,000,000,000,000,000,000,000,000,000,000,000,000,000,000,000,000,000,000.

The name for this number was made up by a nine-year-old, the nephew of Dr. Edward Kasner, an author of mathematics books. He also came up with the name for an even larger number called a googolplex which he described as one followed by as many zeroes as you can write before your hand got tired. A **googolplex** is one followed by a googol of zeroes or 1×10^{googol}.

Further Investigations

For those interested in large numbers and the American and European naming conventions see:

http://www.ecstaticfuturist.com/MiscInfo/numbers.html

Schwartz, David M. *How Much is a Million?* HarperCollins Children's Books, 1993. This picture book will provide the motivation for several investigations. For example, one line reads:

"If you wanted to count from one to one million … it would take you about 23 days."

Consider the prefixes used for naming numbers: **bi**, **tri**, **quad**, **quin**, **sex**, etc.

Large Number Lists

American

1	thousand	1×10^3	1,000
1	million	1×10^6	1,000,000
1	billion	1×10^9	1,000,000,000
1	trillion	1×10^{12}	1,000,000,000,000
1	quadrillion	1×10^{15}	1,000,000,000,000,000
1	quintillion	1×10^{18}	1,000,000,000,000,000,000
1	sextillion	1×10^{21}	1,000,000,000,000,000,000,000
1	septillion	1×10^{24}	1,000,000,000,000,000,000,000,000
1	octillion	1×10^{27}	1,000,000,000,000,000,000,000,000,000
1	nonillion	1×10^{30}	1,000,000,000,000,000,000,000,000,000,000
1	decillion	1×10^{33}	1,000,000,000,000,000,000,000,000,000,000,000
1	undecillion	1×10^{36}	1,000,000,000,000,000,000,000,000,000,000,000,000
1	duodecillion	1×10^{39}	1,000,000,000,000,000,000,000,000,000,000,000,000,000
1	tredecillion	1×10^{42}	1,000,000,000,000,000,000,000,000,000,000,000,000,000,000
1	quattuordecillion	1×10^{45}	1,000,000,000,000,000,000,000,000,000,000,000,000,000,000,000
1	quindecillion	1×10^{48}	1,000,000,000,000,000,000,000,000,000,000,000,000,000,000,000,000
1	sexdecillion	1×10^{51}	1,000,000,000,000,000,000,000,000,000,000,000,000,000,000,000,000,000
1	septendecillion	1×10^{54}	1,000,000,000,000,000,000,000,000,000,000,000,000,000,000,000,000,000,000
1	octodecillion	1×10^{57}	1,000,000,000,000,000,000,000,000,000,000,000,000,000,000,000,000,000,000,000
1	novemdecillion	1×10^{60}	1,000,000,000,000,000,000,000,000,000,000,000,000,000,000,000,000,000,000,000,000
1	vigintillion	1×10^{63}	1,000,000,000,000,000,000,000,000,000,000,000,000,000,000,000,000,000,000,000,000,000
Googol		1×10^{100}	10,000,000,000,000,000,000,000,000,000,000,000,000,000,000,000,000,000,000,000,000,000,000,000,000,000,000,000,000,000,000,000,000,000,000,000,000,000,000,000,000,000,000,000,000

British

1	thousand	1×10^3	1,000
1	million	1×10^6	1,000,000
1	thousand million	1×10^9	1,000,000,000
1	billion	1×10^{12}	1,000,000,000,000
1	thousand billion	1×10^{15}	1,000,000,000,000,000
1	trillion	1×10^{18}	1,000,000,000,000,000,000
1	thousand trillion	1×10^{21}	1,000,000,000,000,000,000,000
1	quadrillion	1×10^{24}	1,000,000,000,000,000,000,000,000

References: Nelson, R D & Nelson, David (Eds), *The Penguin Dictionary of Mathematics*, Penguin, 1999.

TEACHER BACKGROUND

Metric Paper—1

The term *metric paper* is not really correct. The correct term for this paper is the ISO paper size system. The sizes are based on the metric system. Paper may be bought in the A series, such as the A4 copy paper size, the B series and the C series, used for envelopes.

ISO 216 defines the A series of paper sizes as follows:

- The height divided by the width of all formats is the square root of two (1.4142).
- Format A0 has an area of one square meter.
- Format A1 is A0 cut into two equal pieces; i.e. A1 is as high as A0 is wide and A1 is half as wide as A0 is high.
- All smaller A series formats are defined in the same way by cutting the next larger format in the series parallel to its shorter side into two equal pieces.
- The standardized height and width of the paper formats is a rounded number of millimeters.

Because all pages conform to the height:width ratio of 1: $\sqrt{2}$ or 1:1.4142, the height and width will never be a round number. An A4 sheet is 297 mm by 210 mm rather than 300 mm by 200 mm.

For a comprehensive explanation of the system see:

http://www.cl.cam.ac.uk/~mgk25/iso-paper.html

This system allows for all the various A-sized papers to be cut from the larger pieces.

For example, an A3 sheet may be cut to produce two A4 sheets. An A4 sheet may be cut to form two A5 sheets.

A table of A series paper sizes is shown below:

A0	841 × 1 189	1.413
A1	594 × 841	
A2	420 × 594	
A3	297 × 420	
A4	210 × 297	1.414
A5	148 × 210	
A6	105 × 148	
A7	74 × 105	
A8	52 × 74	
A9	37 × 52	
A10	26 × 37	

Most photocopiers provide special keys for enlarging and reducing copies.

A3 ⟶ A4	71%	$\sqrt{0.5}$
B4 ⟶ A4	84%	$\sqrt{\sqrt{0.5}}$
A4 ⟶ B4	119%	$\sqrt{\sqrt{2}}$
B5 ⟶ A4		
A4 ⟶ A3	141%	$\sqrt{2}$
A5 ⟶ A4		

Messy Math

Metric Paper—1

? The sheet of paper this activity is based on is called A4. It measures 297 mm by 210 mm. The next size up is called A3 and is the same size as two A4 sheets joined together. An A5 sheet is half the size of an A4 sheet. The metric paper system ranges from A0 to A10.

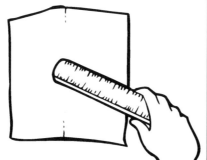

Work out the dimensions of each sheet in the A series. How are they related?

Hints  Ideas

Why doesn't an A4 sheet measure 300 mm by 200 mm?

An A0 sheet has an area of 1 m². Is it 1 m × 1 m? Why?

I wonder how the enlargement button works on the copy machine.

How many A6 sheets fit into an A3 sheet?

What fraction of an A2 sheet is an A8 sheet?

I pressed 2 √ on my calculator and the number 1.414... showed on the display.

I heard that a piece of paper can only be folded seven times.

Notes and Calculations

Questions

Where will you find the information?

What did you find out?

Findings

How will you present your findings?

Teacher Background

Metric Paper—2

Encourage students to work in groups, starting with A4 paper to produce an A3, A2, A1 and A0 sheet of paper. This will involve joining the paper with tape. The A0 sheet should have an area close to $1m^2$.

A5 to A10 sheets may then be produced by folding and cutting.

Then students can measure the length and width of each sheet. If the length is divided by the width, the students will see the result is always close to 1.414. You may need to point out that $\sqrt{2} = 1.414$.

Encourage discussion as to why standardized sized are needed.

Note: when the sheets of paper are laid on the top of each other, a sequence of similar rectangles is formed.

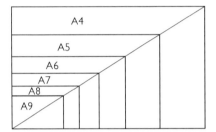

Metric Paper—2

 Shown below is a selection of paper sizes drawn to scale.
What relationships do you notice? Write your answers on a separate sheet of paper.

A1 (841 mm x 594 mm)

A2 (594 mm x 420 mm)

A3 (420 mm x 297 mm)

A4 (297 mm x 210 mm)

A5 (210 mm x 148mm)

A6 (148 x 105)

A7 (105 x 74)

© Didax Educational Resources® – www.didaxinc.com

Messy Math

TEACHER BACKGROUND

Paper, Paper Everywhere

When personal computers were first introduced they were accompanied by promises of a paperless office. The reality has been anything but that! With many people concerned about the environment, the question of how much paper is used in a typical classroom is of concern. Data is readily available by monitoring classroom use, office and library photocopier use.

Paper is usually sold in reams of 500 sheets. A standard sheet measures $8\frac{1}{2}$" x 11". Most reams contain 75 **g/m²** paper—75 grams per square meter. It is probably easier to weigh a ream of paper and divide by 500 to determine the mass of a single sheet of paper rather than calculating from area. Given that collecting data of this nature involves making estimates, it is probably simpler to calculate the answer to the nearest ream.

Further Investigations

Stacking several reams on top of one another, until the stack is as high as a student, will help the class get a feel for larger amounts of paper. Lifting several reams at once will help students appreciate how the mass of paper grows quite quickly. It should also be noted that used paper takes up more space than new, packaged paper.

The price of a ream of paper may be found by examining the price charged at most stationery shops or discount stores. Students may use this information to determine the cost of paper used.

Paper, Paper Everywhere

Plan

 With the introduction of personal computers, the world was promised a paperless office.

The typical U.S. office worker uses about 10,000 sheets of copy paper each year.

Figure out how much paper your class uses in a week, term and year.

How much paper does your class use in a week? _____

How many reams of paper are bought each year? _____

Questions

Where will you find the information?

What did you find out?

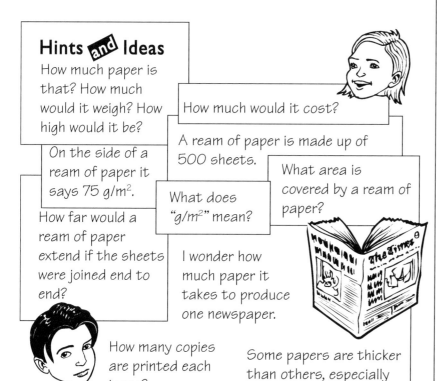

Hints and Ideas

- How much paper is that? How much would it weigh? How high would it be?
- How much would it cost?
- A ream of paper is made up of 500 sheets.
- On the side of a ream of paper it says 75 g/m².
- What does "g/m²" mean?
- What area is covered by a ream of paper?
- How far would a ream of paper extend if the sheets were joined end to end?
- I wonder how much paper it takes to produce one newspaper.
- How many copies are printed each issue?
- Some papers are thicker than others, especially on the weekend.

Findings

Notes and Calculations

How will you present your findings?

Teacher Background

My Legs Hurt

There are several ways a student might tackle this question. For example, he or she could determine his or her average pace length and then keep a log of how many paces walked in a day or week. Keeping a log for a single day may not be representative of the amount of walking done as the timetable for that day may differ significantly from others.

Students could use a range of measuring devices, such as a **trundle wheel** or a **pedometer**, or they might time how long it takes to move from one class to the next. It is important for the students to come to the realization that the *purpose* of the measurement will determine which *measuring tool and units* to use. For example, knowing it is 20 miles from the airport to the city is useful information, but the time taken to drive from the city to the airport is probably more valuable if you are running late to catch a plane!

My Legs Hurt

 Plan

 In a typical school week, children often move around the school to go to different classes (for example, going to the library or the music room).

Questions

How far does an individual student travel in a week, term, or school year?

Hints and Ideas
How much time is spent moving around a school in a typical week, term, or school year?

What is a pedometer?

Where will you find the information?

 How far does a teacher travel during the school day?

How would you design a school to reduce the time lost traveling from one room to another?

Notes and Calculations

What did you find out?

On a separate sheet of paper, explore these challenges.

Challenge
The students in sixth grade are planning to walk along the Bibbulmun Track as part of a school camp. They plan to walk along the track for two days. How far might they travel?

Findings

Hints and Ideas
How far does a person walk in a hour?

How do you think carrying a backpack would affect walking speed?

Do you think a person could keep this pace for eight hours?

How will you present your findings?

Challenge
A 6-year-old boy wandered away from his parents' campsite early one morning. It took four hours to raise the alarm. What search radius would you suggest?

© Didax Educational Resources® – www.didaxinc.com   **Messy Math**

Teacher Background

Popcorn Packets

This activity is designed to confront the misconception that surface area and capacity are linked. This is **not** true—as the popcorn packets demonstrate. It should be noted that children often confuse volume with capacity. *Capacity* may be thought of as the amount a container holds, whereas *volume* represents the amount of space taken up by an object. For example, a refrigerator has a certain capacity, measured in gallons, but takes up a certain amount of space in the kitchen. To add to the confusion, both the words *volume* and *capacity* have different meanings outside of mathematics; e.g. volume is a button on the TV.

Further Investigations

Another common misconception is that perimeter and area are linked. A good problem to try that exposes this misconception involves fixing the perimeter (e.g., the amount of fencing available) and then altering the dimensions to enclose the largest area.

A farmer has 64 m (approx. 70 gd) of chicken wire and wishes to make a chicken yard. How many different-sized yards can be made? Which dimensions produce the yard with the largest area?

Popcorn Packets

Plan

The sixth grade students at Seatown School are planning to go on school camp. To raise money, they decide to sell popcorn at recess. The popcorn containers are to be made from a single sheet of copy paper. The students decided to make cylindrical containers.

Compare the two cylinders that may be made from a sheet of $8\frac{1}{2}"\times 11"$ paper. Fill the cylinders with popcorn. You can use other paper to make the bottom of your container. Which popcorn packet will you use when selling popcorn?

Choice: _____

Explain the reasons for your choice.

Hints and Ideas

Both containers must have the same surface area.

How does the capacity compare?

Notes and Calculations

Questions

Where will you find the information?

What did you find out?

Findings

How will you present your findings?

Messy Math

TEACHER BACKGROUND

Measure Matters—1

A horse is measured from the ground to its shoulderblades. The hand measure was originally the width of an adult hand. Today it is the equivalent of 4" or 10.16 cm. Another old measure used in association with horses is the furlong. Originally the furlong referred to a furrow length—the distance that could be plowed before resting the animal. A furlong is about 200 meters long. Try marking out a distance of 200 meters on the school track. There were exactly eight furlongs to a mile. Most racetracks are basically one-mile ovals and most races are a distance of six furlongs. Along a racetrack you will see poles—each one furlong apart.

Each pole is named by its distance from the finish line. For example, the $1/_8$ pole is one furlong from the finish. The $1/_4$ pole is two furlongs ($2/_8$) from the finish. The $3/_4$ pole is six furlongs ($6/_8$) from the finish. This is the point where the gate will be placed for six-furlong races (the most common distance). The poles are color-coded: $1/_8$ poles are green and whilte, $1/_4$ poles are red and white, $1/_{16}$ poles are black and white.

Definitions

Horsepower is the power needed to lift 33,000 pounds a distance of one foot in one minute (about $1\frac{1}{2}$ times the power an average horse can exert). Used for measuring power of steam engines, etc.

A **league** is a rather indefinite and varying measure, but usually estimated at three miles (4.83 km) in English-speaking countries.

A **fathom** is six feet (1.83 m), the length of rope a man can extend from open arm to open arm.

A **cable length** is the length of a ship's cable, about 600 feet (182.88 m).

A **nautical** mile = 1.1515 miles (1.85 km)

A **knot** is the measure of speed on water, or one nautical mile per hour (1.85 km).

The scattergraph should indicate a strong positive relationship between height and arm span. (See Big Foot on page 16.)

Measure Matters—1

Horsepower

While most countries use the metric measurement system, there are still some measurements and expressions we use that relate to the past. For example, horses are measured in "hands." Originally, the hand was closed and a measure was taken across its width.

Collect data from your class to produce a mean hand size for your class.

Mean hand size = _____

Hints and Ideas

- How is a hand defined today?
- What is a furlong?
- Apparently, racetracks were once measured in furlongs.
- How do you actually measure a horse?
- I wonder what the expression "horsepower" means.

The Sea

You may have heard of the book, *20,000 Leagues Under the Sea*, by Jules Verne. What was a *league*? There are several other nautical expressions based on old measures. For example, the *fathom* was used to measure the depth of the sea. A sailor would lower a weighted rope over the side of the ship, wait until it reached the bottom, then haul and measure it from the fingertips across his outstretched arms.

On a separate sheet of paper, draw a scattergraph and plot each student's height and arm span. What do you notice?

Hints and Ideas

- What is a nautical mile?
- What is a knot?

Notes and Calculations

Plan
- Questions
- Findings
- How will you present your findings?

Plan
- Questions
- Findings
- How will you present your findings?

TEACHER BACKGROUND

Measure Matters—2

Definitions

span A span is derived from the distance between the end of the thumb and the end of the little finger when both are outstretched. It is approximately 9" or 22.86 cm.

cubit A cubit is derived from the distance between the elbow and tip of the middle finger. It is approximately 18" or 45.72 cm.

Measure Matters—2

Body Measures

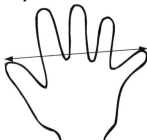

Several old measurement units were based on the body. A *span* was defined as the distance from the tip of the little finger to the thumb when the fingers were spread apart.

Collect some data to produce a mean span for your class.

Mean class span = _____

Our Findings

Builders in ancient Egypt used a measure called a *cubit*. The cubit was defined as the distance from the elbow to the outstretched middle finger.

Collect data to determine a mean cubit for your class.

Mean class cubit = _____

> I would rather buy my material from a person with a large hand.

Our Findings

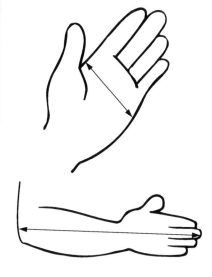

Several other body parts were used to measure. These included the *palm*—the width across the four fingers of the hand. Another measurement called the *hand* was based on the distance across all four fingers and the thumb. The *finger* or *digit*—the width across the index finger at the first joint—was another measure based on the hand. Choose one of these measures to find out which class member has the longest and shortest length. (Remember to measure to the nearest mm.)

Record your results on a separate sheet of paper.

TEACHER BACKGROUND

Measure Matters—3

Foot Length

Have each student compare his or her foot length to the distance from his or her elbow to wrist and note any relationships. In most cases, the two lengths will be the same. Encourage the students to look for other relationships. For example, the length of your **thumb** matches the length of your **nose**. Twice the distance around your **wrist** is about the same as the distance around your **neck** and twice the distance around your **neck** is about the distance around your **waist**. (You may need to be sensitive to any students who are obese when making this comparison.) You may wish to point out that men's shirts are sold according to neck size as the manufacturers realise the relationship between body size and neck size. Many students may also be aware that **arm span** and **height** are roughly the same size. Students might like to find the record for the tallest man and use paper tape to show his height and armspan.

Teaspoons and Tablespoons

In kitchen supply stores you can purchase spoon measures that are calibrated to be used as measuring devices. For example $\frac{1}{4}$ teaspoon is 1.25 mL, $\frac{1}{2}$ teaspoon is 2.5 mL, 1 teaspoon is 5 mL and a tablespoon is 20 mL. If you look on the side of a medicine bottle, dosages are stated in mL and liquid medicine usually comes with a cup measure. Dosage rates tend to vary by age or weight. Recipes on the other hand simply state, "add a teaspoon of salt" or a "tablespoon of honey." You could ask your students why in the case of medicine the dosage is stated in mL but in recipes teaspoons or tablespoons are used. Also, the issue of heaping teaspoons and flat teaspoons could be discussed.

Further Investigations

As a result of completing the various activities involving non-uniform measurement techniques, students should develop an understanding of why a common system of measurement was designed. Have the students investigate the development of the metric system. Many of the terms we use in everyday speech are related to measurement. Think of all the expressions we use that are associated with time; for example, saving time, wasting time and so on. The expression "in a jiffy," while meaning that you won't be long, has a literal meaning—a *jiffy* is $\frac{1}{100}$th of a second.

Measure Matters—3

? The *foot* was another body unit used to measure distances. It was defined as the length from the heel to the toe. Complete a table of everyone's foot length.

What do you notice? _____
_____

Why do you think body measures were abandoned in favor of a standard measurement system? _____
_____
_____

While body measures were convenient, people soon realized that the measures varied from person to person and town to town. Even today, when a recipe calls for the use of a tablespoon, the size differs according to the country in which you live.

Several attempts were made to standardize the measurement system. As far back as the 1100s, King Henry I (1100 - 1135) of England decreed that the *yard* would be the distance from his nose to his thumb when his arm was outstretched.

What was the problem with this standard? _____
_____
_____

In 1215, King John, signed the Magna Carta, which contained numerous laws, including some that pertained to standard measures. Check the Internet or library to see what you can find about measurement systems.

http://www.nsc.gov.au

Hints and Ideas

In Britain, a tablespoon might hold 17.7 mL or 15 mL.

In the U.S.A, a tablespoon only holds 14.2 mL.

In Australia, a tablespoon holds 20 mL. (Big eaters?)

I wonder how much a teaspoon holds.

Name	Foot Length

Our Findings

TEACHER BACKGROUND

Measure Matters—4

The term used to express normal visual acuity (the clarity or sharpness of vision) is "20/20" and is measured at a distance of 20 feet (approx. 6 m). If you have 20/20 vision, you can see clearly at 20 feet what should normally be seen at that distance. If you have 20/100 vision, it means that you must be as close as 20 feet to see what a person with normal vision can see at 100 feet (approx. 30 m).

Having 20/20 vision does not necessarily mean a person has perfect vision—it only indicates the sharpness or clarity of vision at a distance.

carat (c)   A measure of the purity of gold, indicating how many parts out of 24 are pure. For example, 18-carat gold is $3/4$ pure.

and

200 milligrams or 3.086 grains troy. Originally the weight of a seed of the carob tree in the Mediterranean region. Used for weighing precious stones.

pica $1/6$ inch or 12 points

point   .013837 (approximately $1/72$) inch or $1/12$ pica. Used in printing for measuring type size.

yardstick   *Yards* were an early measuring unit. The size of a yard varied. At one point a yard was defined as the distance from the king's nose to the thumb of his outstretched hand. Eventually most villages had a stick or rod (yardstick) that measured three feet—each of which was 12 inches long—that could be used to check the accuracy of the measurement.

peck   A quarter of a bushel. The size of a bushel varied, but the English bushel was around 35 L (approx. 9.5 gallons). For simplicity's sake, a peck would be around 9 L (approx. 2.7 gallons).

Further Investigations

For a comprehensive background to the metric system and a history of measurement see The National Standards Commission website:

http://www.nsc.gov.au

From this website, you may download several leaflets including:

Leaflet No. 6, contains the "History of Measurement."

Leaflet No. 11, contains the "International System of Units."

Measure Matters—4

Plan

? Even though most countries use the metric system, or the Système International d'Unités—SI for short—there are many words, phrases and measures still in use today that are left over from previous measurement systems. For example, we speak of "reaching a milestone" as achieving a goal. This expression dates back to Roman times. When the Roman army marched across a country they would set milestones as a measure of distance travelled.

Several other words and phrases are still used in measurement today. Try to find the origin or an explanation for each.

Questions

a. My Dad has 20/20 vision.

b. How did the metric system come about?

c. Font sizes on the computer are measured in points.

d. What is a pica?

e. The diamond in my Mom's ring is measured in carats.

f. What was a yardstick?

g. What does it mean to inch forward? What is an inchworm?

h. In the tongue twister, "Peter Piper picked a peck of pickled peppers," what is a peck?

Where will you find the information?

What did you find out?

Findings

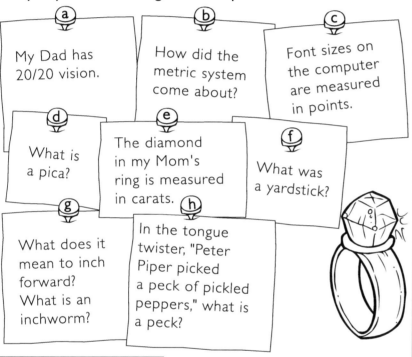

Hints and Ideas

As the Romans marched they would count out 1,000 paces.

In Roman terms this expression for 1,000 paces was mille passus.

Now I see how the word mile came about!

Notes and Calculations

How will you present your findings?

© Didax Educational Resources® – www.didaxinc.com   **Messy Math**

TEACHER BACKGROUND

The Largest Organ in Your Body

The answers from using the different approaches will vary and students may wonder about the accuracy of the various methods.

To find several formulas for calculating Body Surface Area, see:

http://www.halls.md/body-surface-area/refs.htm

Each formula or method will produce a slightly different result. For example, two approaches given on the activity page involve using height, but not weight. Obviously weight is a factor in determining the surface area of the body.

It is crucial for the students to see that a range of answers is acceptable and that in the real world a number of variables affect the calculation.

Further Investigations

There are many related questions that may be considered when exploring this topic, such as the volume–surface area relationship and why babies dehydrate more quickly than adults. Skin cancer and sunburn are also fruitful areas of discussion. Students might also like to explore what the term "third-degree burn" means. Is it serious? How many degrees of burns are there?

The Largest Organ in Your Body

 **Plan**

Questions

? It might surprise you to learn that your skin is the largest organ of your body. Skin protects your body against injury and helps to regulate body temperature. The average adult male has a skin surface area of around 1.8 m² and the average female, 1.6 m².

Use newspaper to make a surface area that is either 1.8 m² or 1.6 m².

There are several ways to calculate your own skin surface area. Try them and compare your results.

- One rough measure involves finding the surface area of your hand and then multiplying by 100.
- Another method involves multiplying the surface area of your foot by 100.
- A third method involves multiplying $\frac{3}{5}$ or 0.6 of your height by your height.
- A fourth method involves doubling your height and then multiplying by the circumference of your thigh.

Where will you find the information?

Hints and Ideas

How many cm² in one m²?

_____

A simple way to count part squares involves only counting squares that are a ½ square or larger.

Forget the rest.

A square meter is 100 cm by 100 cm so there would be 10,000 cm² in one m².

You would think your foot would have a larger area than your hand.

Did you know your skin accounts for 12-15% of your body weight?

What did you find out?

Findings

Notes and Calculations

How will you present your findings?

TEACHER BACKGROUND

The Language of Math

Many students have trouble with mathematics because of difficulties understanding and interpreting the language associated with the subject. Language dictionaries are commonplace in most classrooms but it is less common to see mathematical dictionaries used in the same way. Encouraging students to explain the everyday mathematical language and terms used in the classroom also helps to uncover any misconceptions students might have.

Further Investigations

A good follow-up activity would involve students searching through newspapers looking for words, expressions and terms related to mathematics. Students may create posters of related words when a new topic is developed; e.g. words associated with circles.

References

Useful Math Dictionaries:

- Harcourt Math Glossary
  http://www.hbschool.com/glossary/math2/index_temp.html
- Math Thesaurus
  http://thesaurus.maths.org

The Language of Math

Questions

? Have you ever noticed that a lot of unusual words are used in mathematics? This is because many different civilizations have played a role in developing the mathematics we use today.

For example, the word *decade*, which most people apply to a period of ten years, comes from the Greek word *deka* meaning ten. A decade is literally a group of ten things.

Sometimes a word has several meanings, one in mathematics and others elsewhere. For example, the word *volume* means the amount of space an object takes up and loudness and softness.

Where will you find the information?

Sometimes we use very specific words rather than general terms. For example, the word *perimeter* means to measure around. So, we could be asked to find the perimeter of a circle, but generally we are asked to find the circumference of a circle.

What did you find out?

Locate copies of two or three different math dictionaries and note the way certain words are defined: square, rectangle, rhombus, polygon, diamond.

Make your own 26-word math dictionary containing a word for each letter of the alphabet. Make your definitions simple. You may use diagrams.

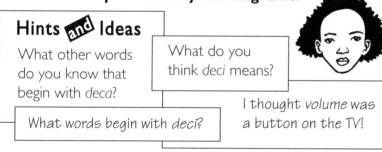

Hints and Ideas
- What other words do you know that begin with *deca*?
- What do you think *deci* means?
- What words begin with *deci*?
- I thought *volume* was a button on the TV!

Findings

Notes and Calculations

How will you present your findings?

TEACHER BACKGROUND

Projections, Enlargements and Distortions

Definitions

dilation: enlarged or reduced

similar: same shape, but not the same size

congruent: same shape and size

This problem is particularly interesting because it illustrates that sometimes a perfect solution to a problem does not exist. It is impossible to represent a sphere (3-D) in two dimensions (map form). The typical map found in most atlases is called a *Mercator Projection*. This projection misrepresents the relative sizes of land masses; hence, North America appears much larger than Africa. The area of the continents is given below. (Note: Figures will vary, depending on the reference used.)

Continent	Area	Approx.
Africa:	30,330,000 km^2	(approx. 12,000,000 mi^2)
Antarctica:	14,250,000 km^2	(approx. 5,500,000 mi^2)
Asia:	44,444,100 km^2	(approx. 17,000,000 mi^2)
Australia:	7,682,300 km^2	(approx. 3,000,000 mi^2)
Europe:	10,531,623 km^2	(approx. 4,000,000 mi^2)
North America:	24,249,000 km^2	(approx. 9,500,000 mi^2)
South America:	17,804,526 km^2	(approx. 7,000,000 mi^2)

The *Peters Projection* represents the relative sizes of land masses, but the shapes of the land masses are distorted. There are several websites that explore this issue in more detail.

Students might like to explore the issue of population density by collecting population data for each country. Many social and environmental issues may then be explored.

1	Russia	6.5 million mi^2
2	Canada	3.8 million mi^2
3	China	3.75 million mi^2
4	United States	3.5 million mi^2
5	Brazil	3.3 million mi^2
6	Australia	2.95 million mi^2
7	India	1.15 million mi^2
8	Argentina	1.05 million mi^2
9	Kazakhstan	1.05 million mi^2
10	Sudan	925,000 mi^2

Different types of grids are located on pages 73, 75 and 77.

Projections, Enlargements and Distortions

? A shape or figure may be distorted, enlarged, or reduced using a grid system.

Reduce the figure shown below by copying it onto the ¹/₅" grid paper.

 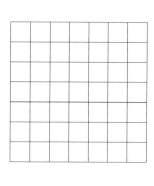 

Notice that the reduced figure is similar to the original figure.

Try copying the figure onto the enlarged and stretched grids (see pages 75-77). What happens to the figure?

_____

Use some of the other grids. The result may remind you of the crazy mirror house that you find at amusement parks and carnivals.

Draw your favorite below.

Hints and Ideas

You might be surprised to find that Africa has an area 1.25 times that of North America.

Look at North America and Africa. Which land mass appears to have the largest area? I wonder why the map looks different.

Take a look at a map of the world and note the size of each of the seven continents.

Try researching under "Mercator Projection" and "Peters Projection" on the internet.

I heard of another map projection called *Buckminster Fuller*.

© Didax Educational Resources® – www.didaxinc.com   **Messy Math**

Projections, Enlargements and Distortions

To be used with the activity on page 73, **Projections, Enlargements and Distortions**.

Projections, Enlargements and Distortions

Projections, Enlargements and Distortions

To be used with the activity on page 73, **Projections, Enlargements and Distortions**.

Projections, Enlargements and Distortions

TEACHER BACKGROUND

Who Erased the Blackboard?

This activity is an example of taking a standard question (e.g. 15 + 17 = ?) and turning it around (? + ? = 32). Providing students with the answer and asking them to supply the question opens up a range of possibilities. Rather than restrict the possibilities, students have many different opportunities to show what they know. For example:

15 + 17 = 32 (whole-number solutions)

$14\frac{1}{2} + 17\frac{1}{2} = 32$ (fraction solutions)

14.5 + 17.5 = 32 (decimal solutions)

-5 + 37 = 32 (negative-number solutions)

Any standard question may be altered this way.

Teaching Point

When students start to mix operations the issue of order of operations will invariably come up. Rather than teach a rule it should be explained that + and − are equally powerful operations. Therefore, you just work the computations from left to right. When multiplication and division are included in a calculation they are completed likewise. However, when addition and multiplication occur in the same calculation, the more powerful operation, multiplication, is done first. Brackets may be used to indicate which part of a calculation should be completed first. Exponents such as squares (e.g. 4^2) and cubes (4^3) are completed before multiplication and division.

Who Erased the Blackboard?

? The teacher walked in as Daniel was cleaning the blackboard. "Oh no!" she cried. "I just finished writing the questions for the math lesson." (All that was left showing on the blackboard was the number 32.) "I know," she said. "You can come up with some new questions! All the questions you write must have an answer of 32."

Write your questions in the box.

32

Hints and Ideas

I'm going to write multiplication (×) and divison (÷) problems.

I'm going to write addition (+) and subtraction (−) problems.

I'm going to write problems that use two operations: e.g. 20 + 20 − 8.

I'm going to include fractions!

What happens if you mix two operations, such as addition (+) and multiplication (×)?

Notes and Calculations

TEACHER BACKGROUND

The Broken Calculator

This activity is very similar to **Who Erased the Blackboard**, on page 80.

This activity tests the students' abilities to work around certain restrictions while performing a calculation. This tests their ingenuity and flexibility with number manipulation as well as their knowledge of number properties. Reducing or increasing the number of restrictions gives the question new life. This question may be used to highlight various options.

The answers students provide will give an insight into their understanding of number manipulation.

The Broken Calculator

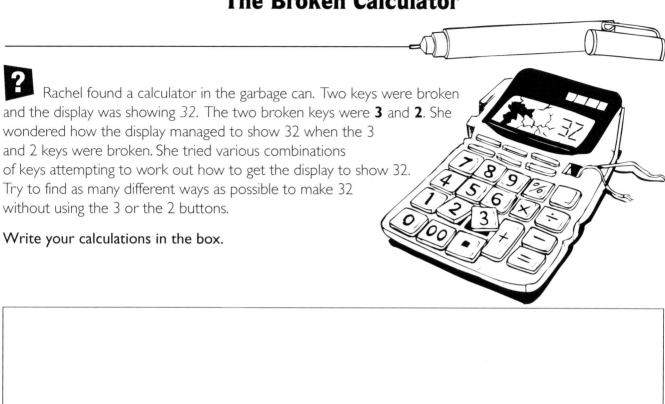

Rachel found a calculator in the garbage can. Two keys were broken and the display was showing 32. The two broken keys were **3** and **2**. She wondered how the display managed to show 32 when the 3 and 2 keys were broken. She tried various combinations of keys attempting to work out how to get the display to show 32. Try to find as many different ways as possible to make 32 without using the 3 or the 2 buttons.

Write your calculations in the box.

32

Hints and Ideas

I used 100 − 68 to make 32.

I used 8 × 4 to make 32.

Notes and Calculations

Where to go from Here?

You have finished all the open-ended mathematics in the book and the students want more. Where do you go from here? There are several possibilities:

- Conduct a brainstorming session with the students. Raise "I wonder" or "What if" questions.

- Look at the world around you with a mathematical eye.

- Take a routine mathematics question and make it more complicated (see **Broken Calculator, Who Erased the Blackboard?**).

Here are few ideas along with brief explanations of some open-ended math. I should explain that I have a family of four boys, including twins; therefore, many of the questions have come about as a result of watching them learn.

- How many hundreds and thousands in a packet of sugar?
  This question came about after knocking a packet over.

- How many hundreds and thousands on a piece of cinnamon toast?

- How many of each color is there in a bag of M&Ms™ or Smarties™ package?
  My children learned their colors quickly when they were allowed to eat the candy afterwards!

  Note: A quick search of the Internet reveals the proportion of each color in a typical bag of candy. This would lead to some data collection and a number of open-ended questions; but best of all, you get to eat the data!

- Design a new Tupperware® toy.
  Just about every family I know has one of those toys in which you push shapes.

You may also wish to explore Fermi problems. Research "Fermi problems" on the internet for some interesting problems to ponder.

NOTES

NOTES

NOTES

NOTES

NOTES

NOTES